FANTASTIC FOSSILS

by Rob Shone

The Rosen Publishing Group, Inc., New York

Published in 2008 by The Rosen Publishing Group, Inc.
29 East 21st Street, New York, NY 10010

First edition, 2008

Designed and produced by
David West Books

Editor: Gail Bushnell

Photo credits:
4t & m, David West; 4b, Joe McDaniel istock; 6b, istock; 45, Bob Ainsworth

Library of Congress Cataloging-in-Publication Data

Shone, Rob.
  Fantastic fossils / by Rob Shone. -- 1st ed.
      p. cm. -- (Graphic discoveries)
  Includes index.
  ISBN-13: 978-1-4042-1088-2 (hardcover)
  ISBN-13: 978-1-4042-9592-6 (6 pack)
  ISBN-13: 978-1-4042-9591-9 (pbk.)
  1. Fossils--Popular works. 2. Paleontology--Popular works. I. Title.
  QE714.3.S54 2007
  560--dc22

                              2007010040

Manufactured in China

CONTENTS

WHAT IS A FOSSIL?

The word fossil comes from the Latin "fossus," which means to dig up. Fossils are the remains of animals or plants that have been petrified (turned to stone).

A trilobite fossil from 450 million years ago.

TYPES OF FOSSIL

Often an animal's remains are eaten, or they rot away so that only the skeleton (or external skeleton, as in certain sea creatures such as trilobites), remains to be fossilized. On rare occasions, an animal may be buried quickly, or by matter that stops it from rotting. This may lead to the formation of a fossil showing details of skin and internal organs. Scientists can learn a great deal about what the animal looked like from these rare fossils. Small insects and pollen grains may be preserved in fossilized resin called amber. These are called resin fossils. Certain plants ooze resin as a form of defense against insects, which get stuck in its sticky mass. Trace fossils are another type of fossil. They are the remains of nests, eggs, droppings, and footprints left by animals. Fossilized droppings, called coprolites, can show what animals like dinosaurs ate.

A close-up of a midge (above) that was caught in the amber (left).

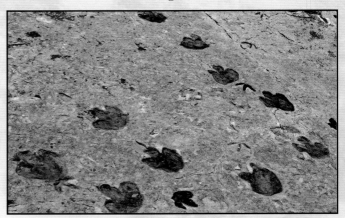

Dinosaur tracks are visible on an exposed layer of Dakota sandstone on Dinosaur Ridge, west of Denver, Colorado.

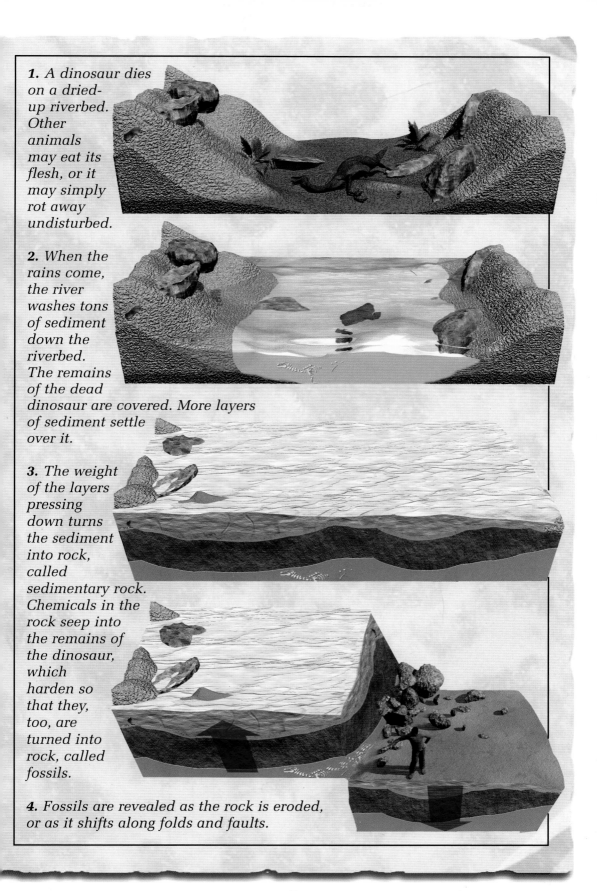

1. *A dinosaur dies on a dried-up riverbed. Other animals may eat its flesh, or it may simply rot away undisturbed.*

2. When the rains come, the river washes tons of sediment down the riverbed. The remains of the dead dinosaur are covered. More layers of sediment settle over it.

3. The weight of the layers pressing down turns the sediment into rock, called sedimentary rock. Chemicals in the rock seep into the remains of the dinosaur, which harden so that they, too, are turned into rock, called fossils.

4. *Fossils are revealed as the rock is eroded, or as it shifts along folds and faults.*

DIGGING FOR FOSSILS

Once a fossil has been discovered, the hard work of removing it from the site begins. Paleontologists, the experts who study fossils, make detailed records of the find before anything is removed.

An early reconstruction of a Triceratops from the 1800s.

FOSSIL COLLECTING

Large fossil finds can take many months to record and remove from the site. The fossils are often cut away along with the surrounding rock, and transported back to the laboratory (see right). Once all the finds have been collected, paleontologists work for months, or even years, separating the fossils from the rock and putting them back together. It takes expert knowledge to do this, especially with finds that involve many pieces. Early fossil collectors often made mistakes. The famous fossil hunter Edward Drinker Cope once put the head of a dinosaur on its tail.

RECONSTRUCTION

When dinosaur fossils were first discovered over 2,000 years ago, people thought they were dragons' bones. Today we can fit the pieces together and reconstruct what early life looked like, from dinosaurs and other prehistoric creatures to our earliest ancestors.

Today's dinosaur museum exhibits can be models that move and roar.

AT THE ARCHAEOLOGICAL DIG SITE

Rock removal
Once the position of the fossil is known, large rocks and top layers of stone and soil are removed. Mechanical diggers, drills, shovels, and sometimes even explosives are used.

Revealing the fossil
Teams of people gradually expose the fossil, using small picks, chisels, and even dentists' tooth picks.

Recording the finds
Photographs, detailed drawings, and maps of the fossil in position are made before anything is removed. These can reveal important clues about how the animal may have died.

Removal
Often the fossils are removed with part of the rock still around them. Small, fragile fossils are sometimes sprayed with glue to keep them together.

Wrapping
Strips of cloth are soaked in liquid plaster and wrapped around the fossils to protect them from damage when they are transported back to the laboratory.

Labeling
All the fossils are labeled before they are put into padded boxes. These are then loaded onto trucks, ready for transport.

MARY ANNING
THE PRINCESS OF PALEONTOLOGY

THE WORLD WAS A VERY DIFFERENT PLACE 200 MILLION YEARS AGO.

MUCH OF THE EARTH WAS COVERED BY WARM OCEANS. MANY CREATURES THAT DO NOT EXIST TODAY LIVED IN THESE SEAS.

AS ANIMALS AND PLANTS DIED, THEIR REMAINS SANK TO THE SEABED, WHERE THEY WERE COVERED BY SILT AND SAND.

OVER TIME, THE SEAS DRIED UP. THE SEABED BECAME ROCK AND THE ROCK BECAME LAND.

1808, LYME REGIS, THE SOUTH COAST OF ENGLAND.

LYME REGIS HAD ONCE BEEN A BUSY PORT, BUT SHIPS HAD GROWN TOO LARGE FOR ITS SMALL HARBOR. HOWEVER, ITS MILD CLIMATE AND PRETTY LOCATION MADE IT POPULAR AS A SEASIDE RESORT.

LYME REGIS HAD ANOTHER ATTRACTION—FOSSILS. TOURISTS SAW THEM AS INTERESTING CURIOSITIES.

LET'S GO IN HERE.

INSIDE.

HERE'S A PRETTY ONE! I CAN'T TELL IF IT'S A STONE OR A SHELL.

THE SHOP WAS OWNED BY RICHARD ANNING. HE LIVED THERE WITH HIS WIFE, MOLLY, AND CHILDREN, JOSEPH AND NINE-YEAR-OLD MARY. RICHARD WAS A FURNITURE MAKER AND, LIKE MANY OTHER PEOPLE IN LYME REGIS, SOLD FOSSILS TO EARN EXTRA MONEY.

IT'S CALLED A SNAKESTONE, MA'AM. IT USED TO BE A SHELL, BUT THEN IT WAS TURNED INTO STONE.

THESE ARE CALLED THUNDERBOLTS...

...AND THIS ONE'S A DEVIL'S TOENAIL.

MY DAUGHTER, MARY, KNOWS AS MUCH ABOUT FOSSILS AS ANYONE IN LYME REGIS.

RICHARD ANNING HAD SHOWN MARY AND HER BROTHER, JOSEPH, HOW AND WHERE TO FIND FOSSILS.

HE HAD ALSO TAUGHT MARY HOW TO CLEAN THEM AND MAKE THEM LOOK ATTRACTIVE TO THE CUSTOMERS.

IN 1810, RICHARD ANNING DIED. HIS FAMILY WAS LEFT OWING A LARGE SUM OF MONEY. LIFE BECAME VERY DIFFICULT FOR THEM. ONE DAY, MARY WAS RETURNING FROM THE CLIFFS WITH A FOSSIL, WHEN...

WHAT IS THAT YOU HAVE, CHILD? LET ME SEE IT.

HMM, A FINE SNAKESTONE. I WILL GIVE YOU HALF A CROWN* FOR IT, AND NOT A PENNY MORE!

*ABOUT ONE DAY'S AVERAGE WAGE IN 1810.

IT WAS THE FIRST FOSSIL THAT MARY SOLD.

LOOK, MOTHER!

WELL! WE SHALL NOT GO HUNGRY **THIS** WEEK! MAYBE YOU SHOULD GO FOSSIL HUNTING MORE OFTEN, MARY.

MARY! OVER HERE! COME AND SEE WHAT I'VE FOUND!

MARY SPENT HER DAYS FINDING FOSSILS FOR HER MOTHER TO SELL. THEY WERE ALWAYS SHORT OF MONEY. IN 1811, SHE WAS SEARCHING THE CLIFFS WITH HER BROTHER...

I THINK IT'S SOME KIND OF CROCODILE.

WE WON'T BE ABLE TO GET IT OUT YET. THE REST IS BURIED TOO DEEPLY IN THE ROCKS.

JOSEPH LOST INTEREST IN THE FOSSIL, BUT MARY DID NOT. SHE WAITED PATIENTLY FOR THE WAVES AND STORMS TO WASH AWAY THE CLIFF HOLDING THE BONES.

BY 1812, MUCH OF THE FOSSIL HAD BEEN FREED. WITH THE HELP OF LOCAL WORKERS, MARY DUG IT ALL OUT.

WELL, I DON'T THINK IT'S A CROCODILE.

MARY'S MOTHER SHOWED THE FOSSIL TO HENRY HOSTE HENLEY, LORD OF THE MANOR OF COLWAY.

I WILL OFFER YOU A FAIR PRICE FOR IT, MRS. ANNING.

TWENTY-THREE POUNDS!* ENOUGH TO KEEP US FOR SIX MONTHS!

*A FARM WORKER EARNED ABOUT 40 POUNDS A YEAR AT THIS TIME.

THE FOSSIL WAS SHOWN AT WILLIAM BULLOCK'S LONDON MUSEUM OF STUFFED ANIMALS.

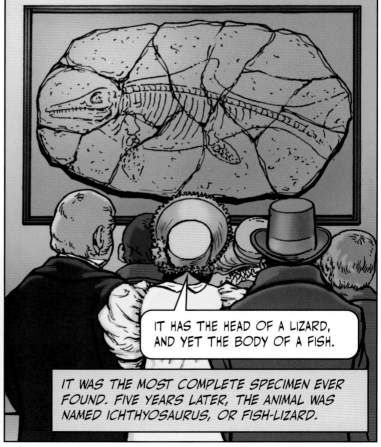

IT HAS THE HEAD OF A LIZARD, AND YET THE BODY OF A FISH.

IT WAS THE MOST COMPLETE SPECIMEN EVER FOUND. FIVE YEARS LATER, THE ANIMAL WAS NAMED ICHTHYOSAURUS, OR FISH-LIZARD.

MARY BEGAN TO STUDY THE FOSSILS SHE COLLECTED MORE CLOSELY. SHE WAS HELPED BY A FRIEND, ELIZABETH PHILPOT, WHO LIVED IN LYME REGIS.

SHE ALSO STUDIED THE ANATOMY OF MODERN ANIMALS, SUCH AS SQUID AND CUTTLEFISH.

IT SAYS HERE THAT THUNDERBOLTS ARE BELEMNITES AND SNAKESTONES ARE REALLY CALLED AMMONITES.

ANOTHER ONE OF HER FRIENDS WAS HENRY DE LA BECHE, WHO WAS INTERESTED IN GEOLOGY.

YOU JUST TAP THE ROCK LIKE THIS...

...AND THERE, ANOTHER AMMONITE.

MARY BEGAN ACTING AS A GUIDE FOR WEALTHY FOSSIL COLLECTORS FROM ALL OVER THE COUNTRY.

WE SHOULD TRY OVER THERE BY BLACK VEN CLIFFS.

MR. DE LA BECHE TELLS ME YOU ARE A PROFESSOR OF GEOLOGY, DR. BUCKLAND.

YES, AT OXFORD UNIVERSITY.

IN FACT, I AM THE UNIVERSITY'S **ONLY** GEOLOGY PROFESSOR. IT IS A NEW SCIENCE, MARY. THE FOSSILS YOU FIND ARE HELPING US TO UNDERSTAND WHAT THE EARTH WAS LIKE MANY YEARS AGO.

MARY ACTED AS A GUIDE FOR DR. BUCKLAND MANY TIMES. SHE FOUND THE FOSSILS, HENRY DE LA BECHE DREW THEM, AND DR. BUCKLAND WROTE ABOUT THEM.

A TRILOBITE! SPLENDID!

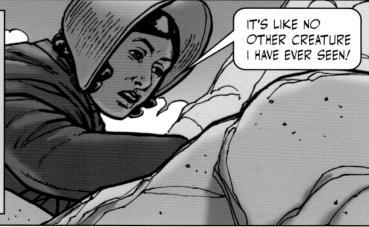

BY 1820, THE ANNING FAMILY FACED POVERTY ONCE MORE. THEY WERE SAVED BY A GIFT OF 400 POUNDS FROM A WEALTHY FOSSIL COLLECTOR. MARY ANNING'S LUCK STARTED TO CHANGE. IN 1823, SHE DISCOVERED THE FOSSIL THAT MADE HER FAMOUS.

IT'S LIKE NO OTHER CREATURE I HAVE EVER SEEN!

MARY SOLD THE FOSSIL TO THE DUKE OF BUCKINGHAM FOR 110 POUNDS. THE DISCOVERY WAS NEW TO SCIENCE AND CAUSED A SENSATION WHEN IT WAS PUT ON DISPLAY. IT WAS CALLED PLESIOSAURUS—ALMOST A LIZARD.

THE FAMOUS FRENCH SCIENTIST GEORGE CUVIER EXAMINED IT.

WHEN I SAW DRAWINGS OF THE FOSSIL, BUCKLAND, I THOUGHT SUCH A CREATURE COULD NEVER HAVE LIVED.

BUT NOW THAT I HAVE SEEN IT I AM AMAZED! IT IS AS IF A SNAKE HAD BEEN THREADED THROUGH THE BODY OF A TURTLE.

NO ONE MENTIONED ITS FINDER, MARY ANNING.

MARY USED THE MONEY TO BUY A HOUSE IN THE CENTER OF LYME REGIS. IN 1826, THE ANNING FOSSIL DEPOT WAS OPENED.

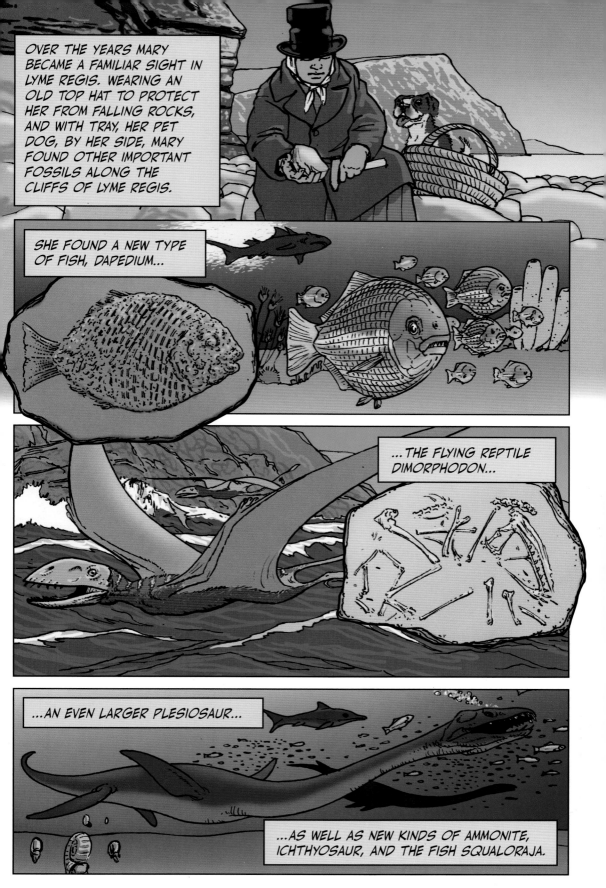

OVER THE YEARS MARY BECAME A FAMILIAR SIGHT IN LYME REGIS. WEARING AN OLD TOP HAT TO PROTECT HER FROM FALLING ROCKS, AND WITH TRAY, HER PET DOG, BY HER SIDE, MARY FOUND OTHER IMPORTANT FOSSILS ALONG THE CLIFFS OF LYME REGIS.

SHE FOUND A NEW TYPE OF FISH, DAPEDIUM...

...THE FLYING REPTILE DIMORPHODON...

...AN EVEN LARGER PLESIOSAUR...

...AS WELL AS NEW KINDS OF AMMONITE, ICHTHYOSAUR, AND THE FISH SQUALORAJA.

MARY MADE OTHER DISCOVERIES...

DR. BUCKLAND, HAVE YOU EVER SEEN ONE OF THESE BEFORE?

I FOUND THIS ONE INSIDE THE BODY OF AN ICHTHYOSAUR FOSSIL.

WHY, YES. IT'S A BEZOAR STONE. I'VE OFTEN SEEN THEM ON THE BEACH.

I THINK BEZOAR STONES ARE CLUMPS OF UNDIGESTED FOOD AND WASTE THAT HAS FOSSILIZED. THEY COULD TELL US WHAT THESE ANIMALS ATE.

BUCKLAND NAMED THEM COPROLITES, OR "DUNG STONES."

ON ANOTHER OCCASION...

ELIZABETH, LOOK AT THIS. I BROKE OPEN A FOSSIL BELEMNITE, AND FOUND A SMALL SPACE FILLED WITH PURPLE POWDER. WHEN I ADDED WATER, IT TURNED INTO INK.

THE ANIMAL HAD AN INK SAC, JUST AS MODERN SQUID DO. IT CAN EVEN BE USED TO WRITE WITH.

18

COLLECTING FOSSILS COULD BE DANGEROUS WORK. ONE DAY IN 1833...

STAY HERE, TRAY.

WHENEVER SHE MADE A LARGE FIND, SHE LEFT HER PET DOG BEHIND WHILE SHE WENT TO GET HELP.

MOMENTS LATER.

TRAY!

THE ROCKFALLS THAT FREED THE FOSSILS COULD ALSO KILL.

MARY ANNING WAS BECOMING WELL KNOWN. SCIENTISTS FROM ALL OVER EUROPE VISITED HER.

BUT THEY GO AWAY, WRITE THEIR BOOKS, AND USE MY KNOWLEDGE AS IF IT WERE THEIR OWN. AND I GET NOTHING, NOT EVEN A "THANK YOU," LET ALONE PAYMENT.

I ENJOY TALKING WITH THEM, HENRY, AND I DO NOT MIND SHARING WHAT I KNOW.

IF IT WERE NOT FOR ME THEY WOULD HAVE VERY LITTLE TO WRITE ABOUT!

IN 1839, PROFESSOR RICHARD OWEN* WENT FOSSIL HUNTING WITH MARY AND BUCKLAND.

MISS ANNING MAY HAVE A NATURAL TALENT FOR FINDING FOSSILS, BUCKLAND...

*OWEN INVENTED THE WORD "DINOSAUR," IN 1842.

...BUT YOU CANNOT COMPARE HER TO MEN OF LEARNING AND SCIENCE FROM ENGLAND'S UNIVERSITIES.

SHE IS AN UNEDUCATED WOMAN OF LOW BIRTH.

ALTHOUGH I AM IMPRESSED BY THE WAY SHE SKIPS OVER THESE SLIPPERY ROCKS.

YOU, SIRS, OUGHT TO SKIP A LITTLE MORE LIVELY YOURSELVES. THE TIDE'S COMING IN!

HEAVENS!

UNEDUCATED? OF LOW BIRTH?

AND NOW ALL THEY THINK I'M GOOD FOR IS CLIMBING OVER WET ROCKS!

MARY SAVED THE TWO SCIENTISTS FROM BEING WASHED AWAY BY LYME BAY'S DANGEROUS TIDE.

20

ENGLAND'S SCIENTISTS MAY NOT HAVE APPRECIATED MARY, BUT ELSEWHERE IN EUROPE SHE WAS SEEN AS A PIONEER IN THE NEW SCIENCE OF PALEONTOLOGY. IN 1844, AN UNEXPECTED VISITOR ARRIVED AT THE ANNING FOSSIL DEPOT.

MISS ANNING, MAY I PRESENT TO YOU HIS MAJESTY, KING FREDERICK AUGUSTUS OF SAXONY.

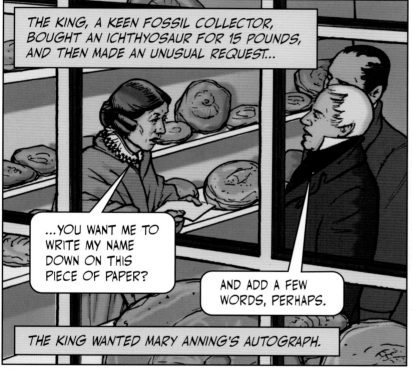

THE KING, A KEEN FOSSIL COLLECTOR, BOUGHT AN ICHTHYOSAUR FOR 15 POUNDS, AND THEN MADE AN UNUSUAL REQUEST...

...YOU WANT ME TO WRITE MY NAME DOWN ON THIS PIECE OF PAPER?

AND ADD A FEW WORDS, PERHAPS.

THE KING WANTED MARY ANNING'S AUTOGRAPH.

IT SAYS "MARY ANNING, I AM WELL KNOWN THROUGHOUT THE WHOLE OF EUROPE."

FOR THE LAST FEW YEARS OF HER LIFE MARY RECEIVED A SMALL PENSION, THANKS TO WILLIAM BUCKLAND. SHE DIED IN 1847, AND WAS BURIED IN ST. MICHAEL'S PARISH CHURCH, LYME REGIS. THE FOLLOWING YEAR HER FRIEND HENRY DE LA BECHE WROTE HER OBITUARY. IT WAS PRINTED IN THE JOURNAL OF THE GEOLOGICAL SOCIETY OF LONDON. IT IS STILL THE ONLY TIME THE JOURNAL EVER PRINTED THE OBITUARY OF A NON-MEMBER.

THE END

21

BONE WARS
THE MARSH & COPE FOSSIL FEUD

1870, THE MUSEUM OF THE PHILADELPHIA ACADEMY OF NATURAL SCIENCES. EDWARD COPE WAS SHOWING HIS RECONSTRUCTION OF THE MARINE REPTILE ELASMOSAURUS TO OTHNIEL MARSH.

IT'S AN IMPRESSIVE DISPLAY, COPE. I COULDN'T HELP NOTICING ONE SMALL THING, THOUGH.

SHOULDN'T ITS HEAD BE AT THE OTHER END OF ITS BODY?

WHAT?

LATER.

IF ONLY IT HAD NOT BEEN MARSH WHO SPOTTED THE MISTAKE, PROFESSOR LEIDY.

THIS WAS NOT THE BEGINNING OF THE FEUD, AND COPE AND MARSH WERE NOT ALWAYS ENEMIES.

MARSH IS RIGHT, COPE. YOU **HAVE** PUT THE HEAD ON THE END OF ITS TAIL.

EDWARD COPE HAD MOVED TO HADDONFIELD, NEW JERSEY, TO BE NEAR A QUARRY WHERE DINOSAUR FOSSILS HAD BEEN FOUND. IN 1886 HE TOOK HIS OLD COLLEGE FRIEND OTHNIEL MARSH TO SEE IT.

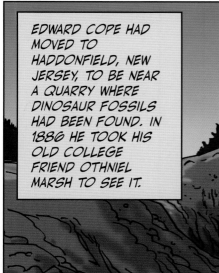

THE QUARRY MANAGERS LET ME KNOW WHENEVER THEY FIND ANY FOSSIL BONES.

EDWARD DRINKER COPE WAS BORN IN 1840 TO A WEALTHY FAMILY FROM PHILADELPHIA. HE SHOWED A GIFT FOR NATURAL HISTORY AT A VERY EARLY AGE. BY THE TIME HE WAS 18 HE HAD BECOME AN EXPERT ON SALAMANDERS. WHEN HE WAS 24 HE WAS MADE A PROFESSOR AT HAVERFORD COLLEGE. COPE HAD A FIERY TEMPER, THOUGH, WHICH MADE HIM MANY ENEMIES.

OTHNIEL CHARLES MARSH WAS BORN IN 1831. AFTER EARNING A DEGREE FROM YALE COLLEGE, HE PERSUADED HIS UNCLE, THE WEALTHY GEORGE PEABODY, TO FOUND A MUSEUM OF NATURAL SCIENCE THERE. IN 1866, MARSH BECAME ITS FIRST PROFESSOR, A POST HE HELD ALL HIS LIFE. WHEN HIS UNCLE DIED, MARSH INHERITED A FORTUNE. MARSH WAS NOT AN EASY PERSON TO LIKE, AND WAS NICKNAMED "THE GREAT DISMAL SWAMP."

I SHOULD LIKE TO MEET THESE QUARRY MANAGERS.

NOT LONG AFTER, COPE STOPPED RECEIVING FOSSILS FROM THE QUARRY.

WE HAD A DEAL!

BUT PROFESSOR MARSH HAS THE CASH!

MARSH HAD PAID THE QUARRY MANAGERS TO LET HIM KNOW FIRST OF ANY FOSSILS THEY FOUND.

MARSH AND COPE TURNED THEIR ATTENTION TO THE NEW LANDS TO THE WEST. EACH YEAR, THROUGHOUT THE 1860S AND 1870S, THE U.S. GOVERNMENT ORGANIZED SEVERAL EXPEDITIONS THERE. MARSH HAD LINKS WITH THOSE LED BY CLARENCE KING AND JOHN WESLEY POWELL. COPE WENT ON THE FERDINAND HAYDEN EXPEDITIONS. THE SURVEYS WERE IMPORTANT TO BOTH MARSH AND COPE. THEY HELPED PAY FOR THEIR FOSSIL HUNTING, AND THROUGH THE SURVEYS THEY COULD PUBLISH THEIR FINDINGS.

IN 1870, MARSH AND A GROUP OF YALE STUDENTS CROSSED THE MISSOURI RIVER INTO NEBRASKA.

AN ARMY ESCORT WENT ALONG FOR THEIR SAFETY. THE JOURNEY TOOK THEM ALL THE WAY TO CALIFORNIA AND BACK.

CONDITIONS WERE HARSH. THEY HAD TO ENDURE BAKING HEAT AND FREEZING COLD...

...FOOD AND WATER SHORTAGES...

PAHH! THE WATER'S BAD. WE CAN'T DRINK IT.

...DEADLY SNAKES...

RATTLERS HAVE GOTTEN AMONG OUR CLOTHES!

...ACCIDENTS...

...AND THEY HAD TO BE ON THE LOOKOUT FOR SIOUX AND CHEYENNE WARRIORS.

FOR THE NEXT 30 YEARS, FOSSIL HUNTERS IN THE WEST EXPERIENCED SIMILAR HARD CONDITIONS.

MARSH DID NOT HAVE THE WEST ALL TO HIMSELF FOR LONG.

IN 1871, COPE ARRIVED AT FORT BRIDGER IN WYOMING. HE WAS ON HIS WAY TO THE BRIDGER BASIN.

WHERE ARE THE MULES AND WAGONS I ASKED FOR, SOLDIER?

PROFESSOR MARSH TOOK THEM. I HAD ORDERS TO LET HIM.

COPE EVEN HAD TO STAY IN THE STABLES WHILE HE WAITED FOR NEW SUPPLIES.

NO FOOD, NO MULES— AND NO BED!

MARSH HAD USED HIS OLD YALE FRIENDS IN THE ARMY TO HELP HIM.

IT WILL BE WEEKS BEFORE COPE IS READY. THAT WILL TEACH HIM TO MUSCLE IN ON MY TERRITORY.

COPE USED TRICKS OF HIS OWN. AT TELEGRAPH OFFICES THROUGHOUT WYOMING...

SO THAT'S WHERE MARSH IS DIGGING.

I'LL LET YOU KNOW IF ANY MORE OF MARSH'S TELEGRAMS COME IN, DR. COPE.

BOTH TEAMS OF COLLECTORS SPIED ON EACH OTHER.

LOOK, SMOKE. WHO DO YOU THINK IT IS, THE SIOUX OR CHEYENNE?

NEITHER. IT'S COPE!

ALL THE TIME MARSH AND COPE SEARCHED FOR FOSSILS. IN KANSAS, MARSH FOUND THE VERY FIRST PTERANODON, A FLYING REPTILE.

IT MUST HAVE HAD A WINGSPAN OF OVER TWENTY FEET!

HE ALSO FOUND FOSSILS OF THE ANCESTOR OF THE HORSE.

A GROWN ANIMAL COULD NOT HAVE BEEN MORE THAN TWO FEET HIGH!

MARSH WAS NOT ABOVE STEALING SKULLS FROM NATIVE BURIAL GROUNDS.

MEANWHILE, COPE WAS COLLECTING FOSSIL FISH...

IT MUST HAVE BEEN NEARLY TWENTY FEET LONG!

...AND MARINE REPTILES.

IT'S A CLIDASTES*.

*A MARINE PREDATOR.

IN 1877, MARSH HEARD OF A LARGE FOSSIL FIND AT MORRISON, COLORADO. MARSH HAD STOPPED GOING INTO THE FIELD AND SENT A TEAM THERE WITH HIS COLLECTOR, BENJAMIN MUDGE.

THE SITE WAS A RICH ONE, BUT THERE WERE PROBLEMS.

THERE ARE PLENTY OF BONES, BUT THEY'RE BRITTLE, MUDGE. THEY'RE BREAKING UP WHEN WE TRY TO DIG THEM OUT.

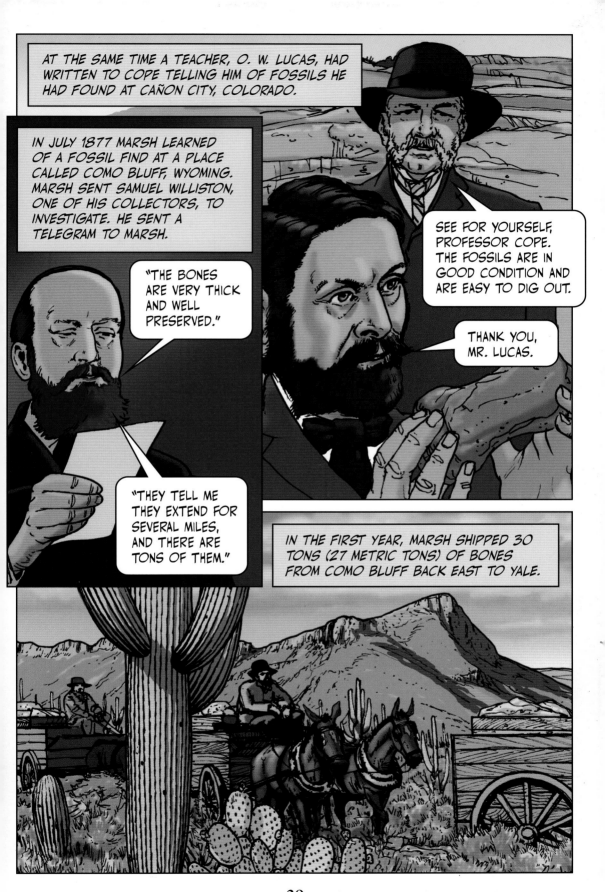

AT THE SAME TIME A TEACHER, O. W. LUCAS, HAD WRITTEN TO COPE TELLING HIM OF FOSSILS HE HAD FOUND AT CAÑON CITY, COLORADO.

IN JULY 1877 MARSH LEARNED OF A FOSSIL FIND AT A PLACE CALLED COMO BLUFF, WYOMING. MARSH SENT SAMUEL WILLISTON, ONE OF HIS COLLECTORS, TO INVESTIGATE. HE SENT A TELEGRAM TO MARSH.

SEE FOR YOURSELF, PROFESSOR COPE. THE FOSSILS ARE IN GOOD CONDITION AND ARE EASY TO DIG OUT.

"THE BONES ARE VERY THICK AND WELL PRESERVED."

THANK YOU, MR. LUCAS.

"THEY TELL ME THEY EXTEND FOR SEVERAL MILES, AND THERE ARE TONS OF THEM."

IN THE FIRST YEAR, MARSH SHIPPED 30 TONS (27 METRIC TONS) OF BONES FROM COMO BLUFF BACK EAST TO YALE.

AS THE FOSSILS WERE SENT EAST, THE DIRTY TRICKS CONTINUED. FOSSIL SITES WERE SPOILED.

THAT'S IT, SPREAD THOSE SCRAP FOSSILS AROUND. IT WILL TAKE COPE MONTHS TO FIGURE OUT WHICH BONES BELONG HERE.

THEY STOLE EACH OTHER'S FOSSILS.

WHEN THEY GET TO PHILADELPHIA, SEND THEM ON TO HADDONFIELD.

I'D LIKE TO SEE MARSH'S FACE WHEN HE FINDS OUT HE'S MISSING A WHOLE BOXCAR'S WORTH OF BONES.

THEY EVEN DESTROYED FOSSILS. NEITHER SIDE WOULD LEAVE BONES FOR THE OTHER TO FIND.

IF WE CAN'T HAVE THEM, NEITHER CAN COPE!

IN SPITE OF THE TRICKS, THE FOSSILS KEPT ROLLING EAST. CRATE AFTER CRATE WAS SENT TO MARSH AT NEW HAVEN AND TO COPE AT HADDONFIELD.

BOTH MEN WORKED FURIOUSLY TO BE THE FIRST TO IDENTIFY A NEW ANIMAL...

...AND WRITE ABOUT IT.

IN 1879, MARSH AND COPE'S PRIVATE WAR MOVED FROM THE WESTERN STATES TO THE EAST, AND WASHINGTON, D.C.

NOT ONLY WAS MARSH A PROFESSOR AT YALE COLLEGE, HE HAD BECOME HEAD OF THE NATIONAL ACADEMY OF SCIENCES.

THE GOVERNMENT WANTS TO SAVE MONEY, MARSH. THREE OR FOUR SURVEYS RUN BY DIFFERENT GOVERNMENT DEPARTMENTS ARE COSTING TOO MUCH. IT COULD STOP FUNDING MY SURVEYS.

DON'T WORRY, KING. AS HEAD OF THE ACADEMY, IT WILL COME TO ME FOR ADVICE.

I WILL RECOMMENDED THAT ONE BODY BE PUT IN CHARGE OF ALL THE DIFFERENT SURVEYS, KING, AND THAT YOU SHOULD RUN IT.

IN MARCH 1879, THE UNITED STATES GEOLOGICAL SURVEY (USGS) WAS FORMED, WITH CLARENCE KING AT ITS HEAD. WHEN MARSH WAS MADE ITS CHIEF PALEONTOLOGIST, HE BECAME THE MOST POWERFUL SCIENTIST IN AMERICA.

COPE HAD NOT BEEN SO FORTUNATE. HE HAD SPENT HIS INHERITANCE, AND WITH MARSH IN CONTROL HAD LOST GOVERNMENT FUNDING. THINGS BECAME WORSE WHEN, IN 1890, THE GOVERNMENT ASKED FOR THE RETURN OF THE FOSSILS HE HAD COLLECTED WHILE ON OFFICIAL SURVEYS.

THIS IS MARSH'S DOING! I SPENT $80,000 OF MY OWN MONEY COLLECTING THOSE FOSSILS! THEY WILL NEVER GET THEM.

MARSH'S VICTORY DID NOT LAST LONG. IN 1892, CONGRESS TRIED TO CUT THE GOVERNMENT BUDGET ONCE MORE. IT SAW THE USGS PALEONTOLOGY DEPARTMENT AS A GOOD PLACE TO START AND ATTACKED MARSH'S WORK ON EARLY TOOTHED BIRDS.

IS THIS WHERE OUR TAXES GO? ON BOOKS ABOUT "BIRDS WITH TEETH"?

IN JULY, POWELL, NOW HEAD OF THE USGS, SENT MARSH A TELEGRAM.

"MONEY CUT OFF. PLEASE SEND YOUR RESIGNATION AT ONCE."

BY 1895, MARSH WAS NO LONGER PRESIDENT OF THE ACADEMY. HE HAD ALSO SPENT ALL OF HIS INHERITANCE.

THERE IS NOTHING ELSE I CAN DO. I SHALL HAVE TO ASK YALE TO PAY ME A SALARY.

IN 1889, COPE WAS MADE PROFESSOR OF GEOLOGY AND PALEONTOLOGY AT THE UNIVERSITY OF PENNSYLVANIA, WHICH GAVE HIM A SALARY. BY THIS TIME HE WAS LIVING WITH WHAT REMAINED OF HIS FOSSIL COLLECTION.

WHEN COPE TOOK MARSH TO SEE THE NEW JERSEY MARL PITS IN 1868, FEWER THAN 20 DINOSAURS WERE KNOWN TO HAVE LIVED IN AMERICA. BY THE END OF THEIR WAR THEY HAD ADDED ANOTHER 130. MANY OF THEIR DINOSAUR FINDS HAVE BECOME WELL-KNOWN NAMES SUCH AS DIPLODOCUS, STEGOSAURUS, AND ALLOSAURUS.

THEY DISCOVERED NOT ONLY DINOSAURS BUT ANCIENT BIRDS, FISH, AND REPTILES...

...AND EARLY NORTH AMERICAN MAMMALS.

COPE DIED IN 1897, MARSH TWO YEARS LATER. WITHOUT THEIR FEUD THEY MAY NEVER HAVE BEEN SO PRODUCTIVE. IN THEIR RUSH TO BEAT EACH OTHER TO NEW FINDS, THOUGH, BOTH MEN MADE CARELESS MISTAKES. IN 1982, THE PEABODY MUSEUM HAD TO CORRECT A DINOSAUR RECONSTRUCTION THAT MARSH HAD MADE. ON THE NECK OF AN APATOSAURUS, MARSH HAD PLACED THE HEAD OF A CAMARASAURUS—A DINOSAUR THAT COPE HAD DISCOVERED.

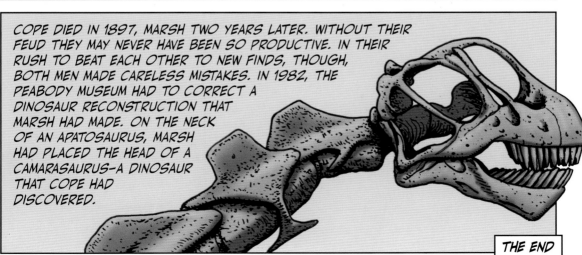

THE END

CSI: SWARTKRANS
A TWO-MILLION-YEAR-OLD MURDER MYSTERY

THE LIMESTONE CAVE WAS ONE OF MANY THAT WERE SCATTERED OVER THE SOUTH AFRICAN PLAIN TWO MILLION YEARS AGO.

OVER HUNDREDS OF THOUSANDS OF YEARS ROCKS AND SOIL, WASHED BY RAINWATER, SLOWLY FILLED IN THE CAVE. ALONG WITH THE DEBRIS WERE THE BONES OF ANIMALS —AND APE-MEN.

IN 1948, PROFESSOR ROBERT BROOM STARTED TO EXCAVATE THE CAVE. ALMOST AT ONCE HE BEGAN TO FIND THE FOSSIL BONES OF AN EARLY HOMINID,* PARANTHROPUS ROBUSTUS. THE FOSSIL EVIDENCE OF HUMANKIND WAS POOR THEN, AND BROOM KNEW THAT THE CAVE WAS IMPORTANT. THE SITE BECAME KNOWN AS THE SWARTKRANS CAVE.

*HOMINID IS THE SCIENTIFIC WORD FOR HUMANS AND THEIR ANCESTORS.

IN 1949, BROOM FOUND PART OF A SKULL, SK54. HE NOTICED THAT IT WAS UNUSUAL.

LOOK HERE. THERE ARE TWO HOLES IN THE BACK OF THIS PARANTHROPUS SKULL.

EVEN THOUGH IT WAS JUST A SKULL FRAGMENT, BROOM COULD TELL THAT IT HAD BELONGED TO A JUVENILE.

THERE'S NO SIGN OF THE BONE HEALING, SO THEY WERE MADE SHORTLY BEFORE OR AFTER ITS DEATH.

THEY LOOK LIKE THEY WERE MADE BY A WEAPON OF SOME SORT.

YOU THINK IT WAS MURDERED?

IT'S HARD TO SAY. IT WOULD SUPPORT THE IDEA THAT THESE APE-MEN WERE HUNTERS AND KILLERS.

IT WOULD BE YEARS BEFORE THE MYSTERY OF THE HOLES WAS SOLVED.

BUT WHO WAS PARANTHROPUS ROBUSTUS? WHAT DID IT LOOK LIKE? WHEN, AND HOW, DID IT LIVE?

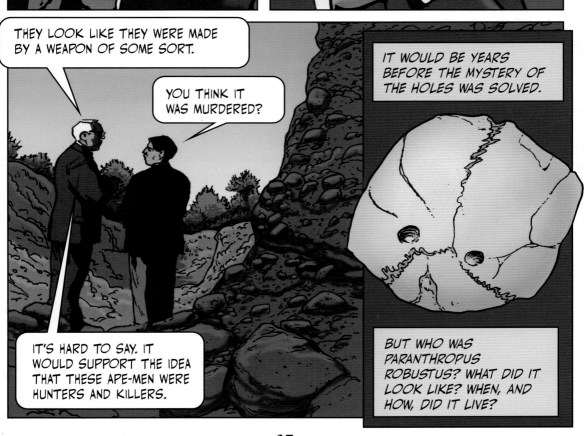

TWO MILLION YEARS AGO THE FORESTS THAT HAD COVERED SOUTH AFRICA WERE ALMOST GONE. GRASSLAND PLAINS AND POCKETS OF WOODLAND HAD TAKEN THEIR PLACE.

THE ANCESTORS OF MANY MODERN-DAY AFRICAN ANIMALS–ELEPHANT, GIRAFFE, AND ANTELOPE–FILLED THIS LANDSCAPE. LIVING AMONG THEM WAS PARANTHROPUS ROBUSTUS.

AN ADULT MALE PARANTHROPUS WAS ABOUT 4 FEET 6 INCHES (1.4 METERS) TALL. FEMALES WERE MUCH SMALLER.

THEY LIVED IN SMALL GROUPS, WANDERING FROM WOOD TO WOOD ACROSS THE GRASSY PLAINS. UNLIKE THEIR RELATIVES, THE APES, PARANTHROPUS COULD WALK UPRIGHT.

WALKING UPRIGHT THROUGH THE TALL GRASS MADE IT EASIER TO SPOT DANGER.

USING THEIR LONG ARMS AND APELIKE HANDS AND FEET, THEY COULD CLIMB THE NEAREST TREE TO ESCAPE FROM HARM.

THEY SPENT MOST OF THEIR TIME FORAGING IN THE WOODS.

PARANTHROPUS HAD MASSIVE JAWS, PERFECT FOR CHEWING TOUGH PLANT STEMS AND ROOTS.

THEY WERE INTELLIGENT ENOUGH TO MAKE SIMPLE WOOD AND BONE TOOLS TO HELP THEM FIND FOOD.

PARANTHROPUS WAS NOT THE ONLY HOMINID IN AFRICA AT THAT TIME. HOMO HABILIS HAD LIVED THERE LONGER, AND WAS MORE ADVANCED THAN ITS COUSIN.

HOMO HABILIS MEANS "HANDYMAN." IT COULD MAKE BETTER TOOLS THAN PARANTHROPUS. HOMO HABILIS COULD SHAPE STONE.

HOMO HABILIS COULD USE FIRE. IT WAS A USEFUL TOOL TO KEEP PREDATORS AWAY.

SO HOW DID THE YOUNG PARANTHROPUS DIE?

COULD ONE OF ITS OWN KIND HAVE KILLED IT? THEY COULD USE TOOLS, AND A TOOL COULD ALSO BECOME A WEAPON.

OR WAS A HOMO HABILIS RESPONSIBLE FOR THE JUVENILE'S DEATH? DID THE TWO SETS OF HOMINIDS FIGHT EACH OTHER?

THEIR TOOLS WERE WELL MADE. TWO BLOWS ON THE BACK OF A HEAD WITH A SHARPENED STONE COULD HAVE MADE THE HOLES.

IN 1965, PROFESSOR C. K. "BOB" BRAIN TOOK OVER THE EXCAVATION WORK AT SWARTKRANS. HE LOOKED AT SK54, THE SKULL WITH THE STRANGE HOLES...

I'VE SEEN MARKS LIKE THIS BEFORE.

BOB BRAIN KNEW HOW THE PARANTHROPUS HAD DIED.

WE WILL NEVER KNOW WHY THE JUVENILE WANDERED AWAY FROM THE GROUP. MAYBE IT SAW SOMETHING GOOD TO EAT. IT GOT NEARER AND NEARER TO THE SWARTKRANS CAVE.

IT DID NOT KNOW THAT THE CAVE WAS ALSO A LAIR.

THE KILLER HAD SEEN ITS YOUNG PREY. IT WAS WAITING FOR THE RIGHT MOMENT.

IT MOVED SILENTLY AND UNSEEN THROUGH THE GRASS, CREEPING CLOSER AND CLOSER TO ITS TARGET.

JUDGING ITS AMBUSH CAREFULLY...

THE LARGE CAT FASTENED ITS TEETH AROUND THE JUVENILE'S THROAT. IT WAS OVER IN SECONDS.

...THE DINOFELIS POUNCED.

THE BIG CAT COULD NOT FEED JUST YET— HYENAS MIGHT STEAL ITS KILL. IT NEEDED SOMEWHERE SAFE TO EAT.

THE DINOFELIS TOOK ITS MEAL INTO TREES NEAR THE CAVE.

WHEN THE CAT CARRIED THE BODY INTO THE TREE, TEETH IN ITS LOWER JAW BIT INTO THE SKULL, MAKING THE HOLES.

EVENTUALLY, THE BONES FELL FROM THE TREE, AND WERE WASHED BY RAINWATER INTO THE CAVE.

BOB BRAIN HAD SEEN THE SAME MARKS ON PRESENT DAY BABOON SKULLS MADE BY LEOPARDS.

THE MYSTERY HAD BEEN SOLVED. OTHER FOSSIL PARANTHROPUS BONES WERE FOUND IN THE CAVE, SHOWING SIGNS OF HAVNG BEEN CHEWED. FAR FROM BEING HUNTERS, THESE EARLY HOMINIDS WERE THE PREY. ONLY PART OF THE SWARTKRANS CAVE HAS BEEN EXCAVATED SO FAR. OTHER MYSTERIES MAY LIE THERE, WAITING TO BE UNEARTHED.

THE END

FAMOUS FOSSIL SITES

There are places around the world where animal remains from the past have been preserved in large numbers, sometimes as well-preserved fossils. These sites are very important because they give clues to the animals' structure and to how they lived. They are called *Lagerstätten*, which is German for "resting place."

A WORLD VIEW

Lagerstätten have provided some of the greatest fossil finds, from Ice Age mammals in the frozen wastes of Siberia to the early hominid fossils in Africa (see map).

Timescales used in map (mya means million years ago)
Precambrian period 4,600–540 mya
Cambrian period 540–500 mya
Triassic period 250–203 mya
Jurassic period 203–135 mya
Tertiary period 65–1.75 mya
Ice Age refers to the last ice age, which peaked at 20,000 years ago and ended about 10,000 years ago.

Burgess Shale
Soft-bodied animals, living in the sea 530 million years ago, have been perfectly fossilized.

Como Bluff
Fossil remains of large Jurassic dinosaurs, such as Apatosaurus and Diplodocus, have been discovered here.

Diplodocus skull

La Brea Tar Pits
Ice Age animals, ranging from mammoths and saber-toothed cats to insects and frogs, were trapped in pools of tar. Over 560 species have been discovered so far.

Eoraptor skull

Valley of the Moon, Argentina
Very early dinosaur fossils, such as Eoraptor from the Late Triassic period, have been found here.

Ichthyosaurus

Lyme Regis
Slate cliffs on the
south coast of
England contain
many fossils from
the Jurassic seas,
such as
Ichthyosaurus.

Solnhofen Limestone
During the Jurassic period,
this was once a shallow
sea with islands. Besides
marine animals, there are
fossils of island animals,
such as Compsognathus
and Archaeopteryx.

Siberia
Thick layers of
permafrost have
trapped and
preserved Ice Age
animals, such as
woolly mammoths.

Mammoth

Archaeopteryx

Protoceratops eggs

Early Hominid
skull

**Chengjiang,
South China**
Soft-bodied sea
creatures from
the early
Cambrian
period have
been perfectly
fossilized.

**Flaming Cliffs,
Mongolia**
Late Cretaceous
dinosaurs, such
as Velociraptor,
Oviraptor, and
Protoceratops have
been found here,
including eggs.

**Olduvai Gorge,
Tanzania**
Fossils of early
hominids and
prehistoric mammals
were found here, by
Louis and Mary Leakey.

Ediacara
Sea creatures from
Precambrian times, such
as sea jellies, sponges,
and worms, have been
preserved as fossils.

Riversleigh
Late Tertiary
fossils of
marsupials such
as kangaroos
have been found,
as well as lions
and crocodiles.

GLOSSARY

academy A society for experts in the same field of study.

amber A transparent yellow fossilized resin that flowed out of coniferous trees.

ambush To come out of hiding and attack by suprise.

anatomy The structure of living things.

classify To put into groups things that share the same features.

curiosity An unusual object.

erode To wear away, or destroy gradually.

expedition A planned journey taken by a group of people.

fossil The remains of plants and animals that have turned to stone.

geology The study of the Earth's structure.

hominid The scientific name for human beings and their animal relatives and ancestors.

inheritance Belongings that are passed from older family members to younger ones.

journal A specialized newspaper or magazine.

laboratory A room or building specially equipped for scientific experiments and research.

marsupial A mammal that spends some time growing in a pouch on its mother's stomach.

obituary A short biography of someone who has just died.

ooze To flow out slowly.

paleontologist Someone who studies fossils.

permafrost A layer of soil below the surface of the ground that stays frozen all the time.

petrify To turn into stone.

preserved Saved from loss, damage, decay, or deterioration.

reconstruction A process of rebuilding something.

resin A thick, waterproof liquid that seeps from some plants and trees.

salamander A lizardlike amphibian, related to frogs and newts.

salary The fixed, yearly amount paid to someone for their work.

sediment Solid material, such as rock fragments, that have settled at the bottom of a sea, lake, or river.

silt Fine sand and mud deposited by flowing water.

specimen A single example of something used for study or display.

structure The way in which the parts of a thing are arranged or organized.

survey An expedition to record and map an area of land.

vanity Extreme personal pride.

FOR MORE INFORMATION

ORGANIZATIONS

The Virtual Fossil Museum
Web site: http://www.fossilmuseum.net
E-mail: webmaster@fossilmuseum.net

Aurora Fossil Museum
400 Main Street
PO Box 352
Aurora, NC 27806-0352
(252) 322-4238
Web site: http://www.aurorafossilmuseum.com
E-mail: aurfosmus@yahoo.com

FURTHER READING

Cefry, Holly. *Fossils*. New York: PowerKids Press, 2003.

DK Eyewitness Books. *Fossil*. New York: DK Publishing, 2004.

Spilsbury, Louise and Richard Spilsbury. *Journal of a Fossil Hunter*. Chicago: Raintree Library, 2006.

Stewart, Melissa. *Fossils*. Minneapolis: Compass Press, 2003.

Wood, Robert Muir. *Discovering Prehistory*. Milwaukee: Gareth Stevens, 2002.

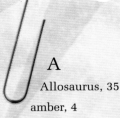

INDEX

Web Sites

Due to the changing nature of Internet links, the Rosen Publishing Group, Inc., has developed an online list of Web sites related to the subject of this book. This site is updated regularly. Please use this link to access the list:

http://www.rosenlinks.com/gd/fossils/